Henry Ling Roth

Notes on Continental Irrigation

Henry Ling Roth

Notes on Continental Irrigation

ISBN/EAN: 9783337396206

Printed in Europe, USA, Canada, Australia, Japan

Cover: Foto ©berggeist007 / pixelio.de

More available books at **www.hansebooks.com**

NOTES

ON

CONTINENTAL IRRIGATION

WITH PLATES.

BY

HENRY LING ROTH.

LONDON:

TRÜBNER & CO.

MDCCCLXXXII.

CONTENTS.

LIST OF PLATES.

CONTINENTAL IRRIGATION.

I. BOUCHES DU RHONE (S. OF FRANCE).

THE rich country lying around the mouths of the
Rhône has, owing to its position and the peculiar
climatological influences which surround it, from
time immemorial suffered severely from excessive
droughts. To the rural population the consequent
distress has been great, for early in the summer all
young vegetation not artifically watered withers up.
The sun withdraws the moisture from the ground
and this moisture is carried away by the "mistral," a
fierce cold wind which rushes down the valley of
the Rhône, leaving the country it traverses dry and
bare. The warm moist air is carried away across
the Mediterranean, and is replaced by a cold dry air
from the north. The cold air as it sweeps along is
gradually warmed, and as it increases in temperature
so does it increase its power of taking up and re-
taining moisture. This now warm moist air being
continually removed and replaced by a colder air,
the desiccating power of the atmosphere is always
fully active. The country being apparently cut off
from rain-bearing winds by the Chaine des Alpines

is not recompensed for these continuous withdrawals, the young crops consequently dry up, and with them the attendant hopes and income of the farmer.

It is now nearly 400 years since the pioneer of modern irrigation, Adam de Craponne, noticing the distress which appeared to occur every seven years out of eight, bethought himself of building an irrigation canal by means of which he would overcome the natural obstacles to successful farming. Until the first Revolution, little or no use was made of the canal, but since then it has become highly prized, and farmers are only too glad to avail themselves of one of their greatest artificial blessings. At present every year witnesses a few new fields irrigated for the first time, and no field once irrigated is ever again allowed to suffer want of water, so that the system is extending slowly but surely. To show how the proximity of the canal, or its conduits, has improved the value of the land, a direct proof of the utility of irrigation works, it may be mentioned that land which previously sold for 1000 to 1200 Francs per Hectare (£16—19 per acre), will, now that it is connected with the canal, fetch as much as 6000 Francs per Hectare (£96 per acre).

Since the building of the Canal de Craponne (80 Kilometres, or 49½ miles, long) several others have been built, of which, in this district, the Canal des Alpines (120 Kilometres, or 74½ miles long) appears to be the most important. The Canal de Craponne

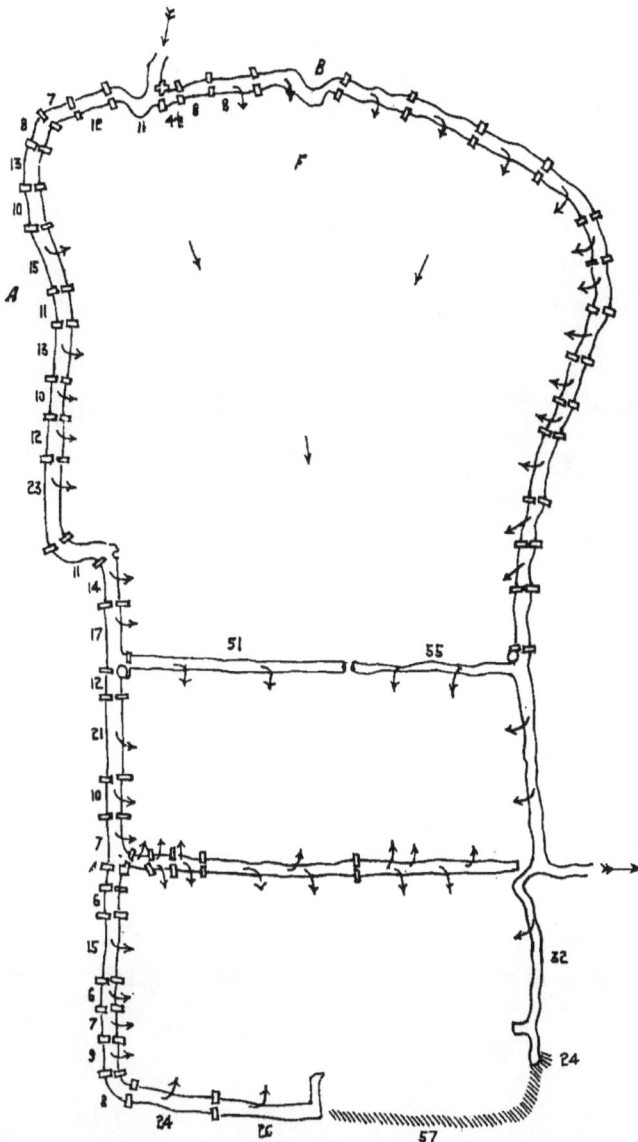

Fig 1

Plan of an irrigated meadow near Arles
Depth of distributor at A 2½ ft., ditto at B 2 & 2½ ft
Slope of the field at F 0'02
The distances between the sluices varied with the slope of the distributor
The figures indicate the distances between the sluices in metres
The arrows indicate the slope of the field, & the direction of the water

is about 2 metres deep by 4 metres broad (6½ x 13 feet); it is in part lined with stone, and was so probably throughout at first. The sluices chiefly in the form of iron-plates are set in stone or solid masonry. In technical terms the Craponne Canal consists of three distinct canals, as it has three different slopes; according to M. Pareto, they are as follows :—0·00086, 0·0004 and 0·0023. It carries upwards of 12,000 Litres (424 cubic feet) per second, and is calculated capable of irrigating 12,000 Hectares (29,600 acres); as, however, all the adjoining lands are not irrigated yet, it supplies water to about 8,000 Hectares (20,000 acres) only. The Canal des Alpines is broader and deeper, and its slopes vary more; it consists of seven distinct canals, the slopes being 0·00008, 0·002, 0·0023, 0·0003, 0·0004 and 0·0005. This Canal taps the river Durance some miles to the west of the weir, which diverts part of the river water into the Canal de Craponne. It is, however, only permitted to withdraw water down to a certain depth, so as not to inconvenience the concessionaires of earlier built canals; it also remains empty during the winter as its overflow opens into a " canal de dessèchement " (large open drain), which being generally full in winter cannot receive the surplus water from this Canal des Alpines. This canal is fitted throughout with double-sluices, by which means the water runs into a small chamber a few feet square, and its outflow is thereby better controlled. The Craponne

Canal has flour mills throughout; the Alpines Canal, owing to want of water in winter, has to do without them. These mills rather increase the velocity of the water, as a fall of several feet is requisite at every mill. The Durance, which supplies these canals and several others, is a shallow, broad, and rapid river, whose waters are generally turbid. It conflues with the Rhône.

From the Canal proper conduits (main, chief, or primary distributors) run out at irregular intervals in accordance with the requirements of the adjoining lands. From these conduits again secondary and lesser distributors branch out, and so on. Occasionally a secondary distributor runs out from the canal and round the field it irrigates; fields are always irrigated from the smaller distributors, never from the canal or conduit direct.

Plate II

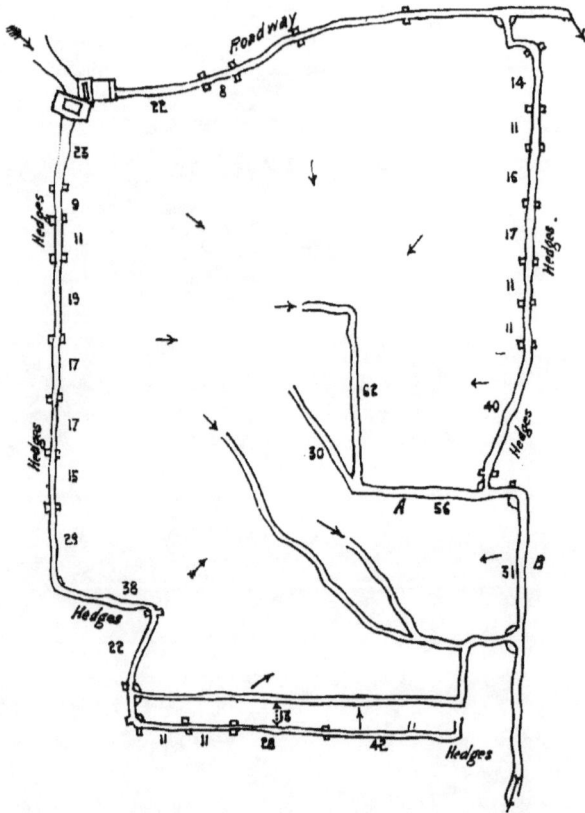

Fig 2

Rough plan of meadow near Burzacotte naturally irrigated

The figures denote distances between the sluices in metres

At A the depth of the drain is nearly one metre

At B the distributor acts alternatldly as an irrigating conduit, & a drain

1. MEADOW IRRIGATION.

THE system for the irrigation of meadows chiefly
followed near Arles is that known as semi-submer-
sion par deversement which is the most simple as
well as probably the most primitive. It is carried
out as explained by Fig. I. The distributors run-
ning round the field are partitioned by roughly con-
structed "vannes" (primitive sluices, see Fig. III).

When an irrigation is to be carried out the irrigator
closes the first sluice by means of a board which is
retained in its position by the pressure of the water,
and the water not being able to pass collects there.
Now the inner-side, or field-side, of the distributor is
lower than the outside, hence when the distributor
is full the water over-flows on to the meadow until
it nearly reaches the bottom, or, in cases where the
meadow is irrigated from all sides, until it has nearly
reached the lowest part; the irrigator then stops the
flow by removing the board to the next "vanne"
where the same performance is repeated and so on
until the whole field has been watered. The water
is turned off by removal of the board before it has
reached the farthest corner, because experience has
taught that when it has reached a certain point,
varying with the slope of the meadow (or slope of
the bed on arable land), it will continue to flow on
to the end without a further addition of water; if
the water were not turned off when it has reached
that point there would be too great an accumulation

of it there, and the grass would grow rank and swampy. Occasionally, chiefly on meadows where otherwise impracticable, but never on arable land, the water is allowed to run into an overflow which forms a distributor for a lower field (much like terrace irrigation) see Figs. II. and XI. Irrigation is generally commenced at the extreme end of the distributor on a field, and is worked backwards. The peasants appear to find this method more convenient than beginning at the upper end.

The soil is for the most part alluvial, originally rather full of stones said to have been brought down by successive floods, and is of a light clayey and sandy nature; the depth varies from 15 to 90 centimetres (6 to 35 inches). The sub-soil is gravel chiefly.

In preparing the land, at the commencement almost wholly wild, the stones were removed, slight hollows filled up and small eminences levelled down; of this, however, there does not appear to have been much to do, as the land was originally well adapted for irrigation. Generally the edges of the meadows were slightly raised, but this is not a necessity. In most cases the peasants laid out their own little plan without any engineer's help and, as they knew nothing about taking levels, furrows were cut along the apparently highest points, from actual trials by means of these with water it was determined where the distributors were to run. (It was, of course, only for the smaller dis-

Plate III

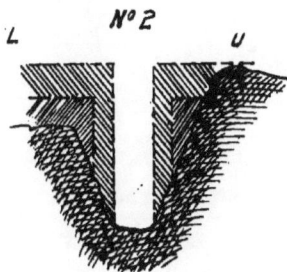

N°2

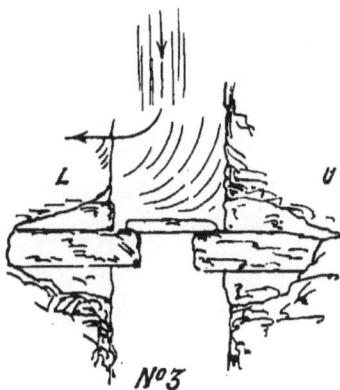

N°3

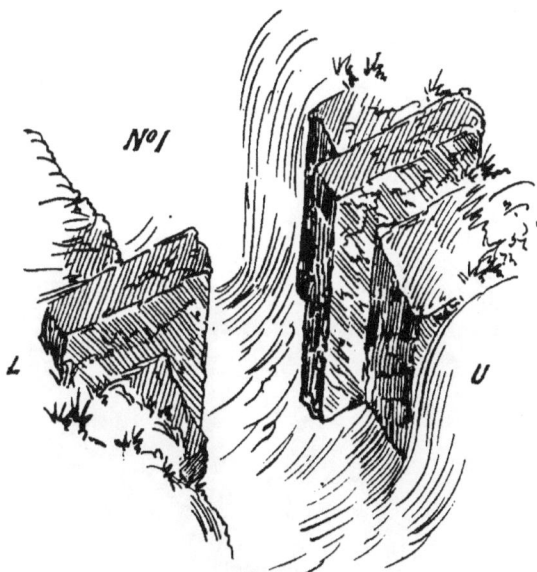

N°1

Fig 3
Sluices in use at Arles

U   Upper side } of distributor
L   Lower side }

tributors that this was done, as the canal Company laid out and built all the conduits and other large distributors). The positions of the distributors having been settled, the next thing to do was to regulate the distances apart of the "vannes." The higher the part of the field, that is the greater the slope and the deeper the distributors (which latter vary from 6 inches to 3 feet in depth) the closer together must the vannes be situated—there is, however, no fixed rule; the widths of the distributors increase with decreasing depth, but here also no fixed rule is adhered to. All this the peasants had to learn for themselves, for instance, in determining the position of the vannes they were guided by practice as follows: suppose two vannes were 7 metres apart originally, for the sake of a trial; well when the water is let into the distributor it is found that owing to the fall of the distribution the water accumulates at the lower vanne and consequently the whole length of that part of the meadow touching the distributor thus cut off is irregularly irrigated, hence the lower vanne must be placed nearer to the upper one, say at a distance of 5 metres. If on the contrary the original distance is found to work well, a trial is made to see whether it cannot be increased say to 8 or 9 metres, as it naturally follows that the further apart the vannes the less work and care is required in keeping the distributor in order and in superintending the irrigation. The distributors when not dug out of

the ground, but built above in order as occasion requires to cross lower ground, are made of earthwork
and turfed.   Thus in Fig. IV. where the distributor
A B has to cross the hollow C E D the peasants
would raise a small canal of earth-work; the earth
of which the canal is to be built, however, is not
merely thrown down and made into proper shape,
but that portion of the hollow it has to cross is dug
up to a depth of 12 to 18 inches, so that the earth-
work and the foundation combine more effectually
to prevent any leakage.

Agricultural operations began by breaking up
the land (to what depth unknown) when lucerne
was sown.   Thanks to irrigation and to a quantity of
stable manure the lucerne grew wonderfully for
four years, generally giving five to seven cuts annually.   After lucerne grasses were sown and the
land remained meadow ever since.   These meadows
give three good cuts per season (the first about end
of May, the second about commencement of July,
the third about end of September), and a winter
crop which is eaten off by cattle, which are useful
to the land by their droppings.   These meadows
are manured generally in November once every two
years, with 250 Francs worth of farm-yard manure
per hectare, valued at 6 Francs per cubic metre,
(5s. per 35 cubic feet) or 8 (?) loads to the
acre.   The irrigation season lasts from 1st of April
to 30th September, the drought commencing on or
about the fifteenth of the former month.   Kitchen

Plate IV

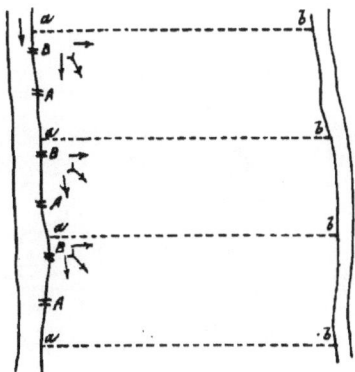

Fig 6

Wheatfield near Mollèges
prepared for irrigation.
→ indicate the slope of the land
⇢ indicate the resultant slope
of the land

Sketch to illustrate preparation of land for main distributor

Fig 4

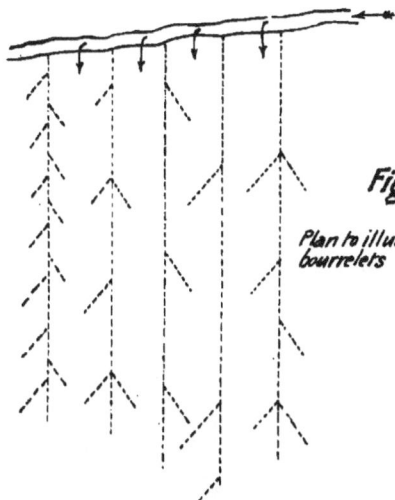

Fig 7

Plan to illustrate branched
bourrelets

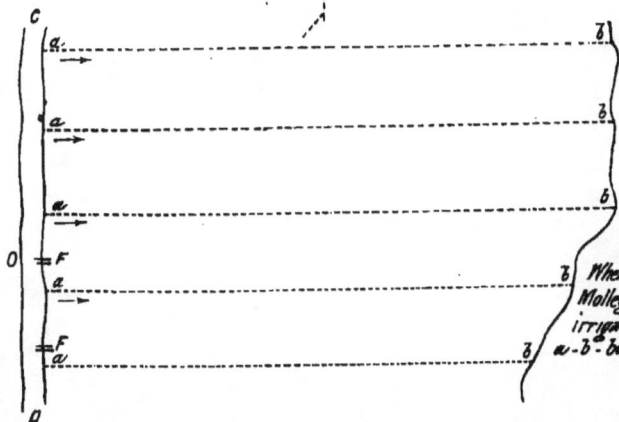

Fig 5

Wheatfield near
Mollèges, prepared for
irrigation
a - b - bourrelets

gardens are watered every 3 or 4 days; meadows 3 times a month and other fields once or twice a season.

The amount of water considered necessary to irrigate a hectare is calculated to cover the surface of the meadow land to a depth of six centimetres, thus the amount of water given to a hectare is 600 cubic metres (21,680 cubic feet per acre) per irrigation.

The Canal de Craponne Company know the sizes of all the fields, and charge from 25 to 40 Francs per hectare per season (8s. 2d. to 13s. per acre).

The laborer who attends is paid 3 to 3½ Francs (2s. 6d. to 3s.) per day, and in that time he is *said* to be able to irrigate 10 to 12 hectares (25 to 30 acres). The Company only carry the water to the fields, the owners having to build the distributors which surround (and occasionally intersect) the fields; to make these and put in the " vannes " involves an expenditure of about 50 Francs per hectare (16s. 4d. per acre), not including the labor of the proprietors, who with their families do the work.

The fields are for the most part very small, varying from less than one up to three hectares (under 2 up to 7½ acres), and the holdings are often not much larger.

2. Irrigation of Arable Land.

The irrigation of arable land is more practised in the neighborhood of St. Remy and Mollèges than elsewhere. Although it has made more progress here than at Arles, it cannot be considered other than still in its infancy. In the neighborhood of Arles there are several peasants who grow wheat without irrigation, notwithstanding that they are benefitted by irrigating their meadows and gardens. It seems this meadow business is so lucrative that no one cares much to attempt anything else which may turn out better or worse. Cereals, not irrigated, yield little or nothing; they flourish up to the middle or end of April, after which, owing to the drought, they make little progress unless an exceptional rain happen to fall. As the return is very poor when irrigation is not resorted to, the peasants are afraid to use manure which, of course, is another cause of so small a return.

It is owing chiefly to the efforts of the Canal des Alpines Company that a stimulus has been given to the growth of cereals, almost wholly wheat, by showing the peasants how profitable wheat crops become when irrigated. According to M. Cacheux, irrigation increases the crops by $4\frac{1}{2}$ hectolitres per hectare (5 bushels per acre), and according to the peasants whose figures are, however, generally vague, it doubles the produce, giving $3\frac{1}{2}$ quarters per acre if the land is manured as well. The average

irrigated crop gives 120 doubles (24 hectolitres) per
hectare (3½ qtrs. per acre) and @ 22 Francs, average
price per hectolitre (53s. per qtr.!) means a gross
return of 528 Francs per hectare (178s. per acre).
The peasants pay 25 Francs per hectare per season,
and repairs are calculated at 7 to 9 Francs per
hectare; both figures much the same at Arles. The
land belongs, with very few exceptions, to the
peasant who cultivates it, and he is generally aided
by his children or relations; when emergencies
occur neighbors are always ready to turn out and
render assistance. The item 7 to 9 Francs for
annual repairs arises as follows : each commune has
a syndicate which carries out all the general repairs
on the secondary and smaller distributors (with
which the Canal Company have nothing to do), and
which amount to 4 or 5 Francs per hectare, the
remaining 3 to 4 Francs are what it is said to cost
the peasant to get his field in order.

Arable irrigation is carried out as explained by
Figs. V. and VI., as follows :—The land is generally
laid out, cleared of stones, and cleaned as described
under meadow irrigation, but instead of following
the natural slopes of the ground, it is levelled down
to one slope (as in Fig. V), or two slopes (as in Fig.
VI.) When this has been done *bourrelets*, or long
narrow ridges are drawn along the length of the
field in the direction of the slope, at various distances
apart; the space between is called a plane. These
bourrelets serve to regulate the flow of the water

B

over the field, and help it to overcome these very
small irregularities of the surface which are un-
avoidable.   These *bourrelets* are destroyed with
ploughing up and renewed every season.

On the wheat-field, Fig. V., at Mollèges, the planes
were 7 metres (22 feet) broad and 70 metres (76
yards) long.   The water arrives in the distributor,
C D, finds its passage barred at O, and naturally
enters the field at F F whence it spreads laterally
as far as the *bourrelets* permit, and flows on unin-
terruptedly until it reaches the end of the plane.
On the wheat-field, Fig. VI., the *bourrelets* were 10
metres (32 feet) apart, which distance, a peasant
informed us, was too great for the slope of the land,
the water not spreading uniformly; the same peasant
also pointed out another fault; the water was
allowed to enter at A instead of at B, which was at
a higher level.

Narrow planes are generally preferred, that is to
say such as vary from 2 to 5 metres (6½ to 16¼ feet) in
breadth, hence the *bourrelets* are generally at those
distances apart ; everything depends on the slope of
the land.   The *bourrelets* are about 5 to 6 inches
high.   Three metres (9 feet 9 inches) may be taken
as a maximum breadth for planes on fields which are
considered well irrigated.   They are not made longer
than 150 metres (162 yards), though this length is
probably due more to the smallness of the fields
than to any practical obstructions to the flow of
water.   There was, however, one exception where

the length of the plane reached 200 metres (nearly 10 chains).

The slope in a distributor here was *said* to be 0·002. On a plane 80 metres (nearly four chains) long the water took an hour to reach the bottom the first time, and 20 minutes less the second and third time; on another plane in an adjoining field, 65 metres (68 yards) long the water required 40 minutes to reach the bottom the first time and about 10 minutes less the second and third time. Our informer told us that the slope of the latter plane was 0·004. The former plane was 7 metres (7½ yards) broad (which as we were informed is too great a breadth), this gives an area of 5·6 ares (670 square yards)—to water this one hour is required, or for an hectare under similar conditions nearly 18 hours are necessary (nearly 7 hours per acre). The peasants told us they could irrigate 1 to 1½ hectares (2½ to 3¾ acres) per day. With heavy slopes the water runs along quicker than on more level surfaces; on big slopes more water is also required, and it is allowed to run for a longer period as there is less absorption when water runs over the surface quickly. Heavy slopes are unsuitable on arable land, as there is too much soil displaced by the water, which displacement or wash should never take place. Generally speaking at least two planes can be irrigated at once, so that actually two to three and more acres can be irrigated in seven hours.

в*

On arable land there is less leaf-surface for trans-
piration than on meadow land where the grasses
grow very thickly, consequently less water is required so far as the crop itself is concerned; but
arable land being more exposed to the drying action
of sun and wind, this loss by evaporation has to
be compensated for. In practice arable lands under
cereal crops receive the same quantity of water per
single irrigation as meadows do, but they are not
irrigated so often as the latter. The Canal des
Alpines Company gives 30 litres per second for 6
hours per single irrigation for arable land, that is to
say it gives 648,000 litres per hectare (9,265 cubic
feet per acre) per single irrigation. Irrigation
commences on or about 15th May on this Canal
and is repeated for careals once, rarely twice, during
the season. Subsequent irrigations, if required,
after the first in the season are given after the formation of the cars and before flowering and then
between flowering and the formation of the grain,
but never while the crop is in flower. It is found
that the plant when in flower is too tender to be
able to withstand irrigation. The peasants irrigate
in the morning or evening; the morning is considered the most suitable, the soil being then
colder than the water. Cold water allowed to flow
on to soil of which the temperature is higher, chills
the plants and checks their growth for a time.

Wheat ripens towards the middle or end of June;
the atmospheric heat being then still on the increase,

irrigation for meadows is carried on until about October.

The water should always be made to flow with regularity, that is to say one part of the field must receive as much water as any other part, there should nowhere be either excess or want; if there be excess the water has been wasted, and if insufficient water be given time and labour have been wasted, and the crop will turn out uneven and patchy. This regularity depends chiefly on the evenness of the slope, and on the land being free from weeds.

Vineyards are irrigated in winter by being completely submerged, the water remaining on the soil a considerable time.

The waters of the Durance, which supplies these two Canals (besides others), bring down annually large quantities of ooze which fertilise the soil irrigated, and which naturally to a certain extent share in the cause of the success which has attended irrigation throughout this part of France. It has been computed by M. Hervé Magnon that the Durance carries down annually 1,100,000 cubic metres of this ooze equal to about 100,000 tons of guano. In some localities advantage of this is taken to increase the soil in a field; the water is let in on to the field and allowed to stand until all matter held in suspension has subsided, when the water is run off and a fresh quantity allowed to flow in, and so on until the field is well fertilised or until a barren field has been made fertile. At Tarrascon we saw two such fields.

II. VALENTIA, GRENADA AND SEVILLE (Spain).

Irrigation of Arable Land.

In Spain the extreme nature of the climate is generally considered to be due to the denudation of the country of its forests.   Alternate floods and droughts are the obstacles to good cultivation that agriculturists have to battle with there.   The country is intersected by a large network of irrigation canals, without which there would be no agriculture to speak of.   At Valentia, for instance, all the water from the River Guadalaviar is diverted into the canals so that at the town one can walk dry-footed from bank to bank.   Here too the average fall of rain during the hottest months of the year is about $4\frac{1}{2}$ inches, so that irrigation works in this locality are absolutely necessary.

We here witnessed a field of wheat undergoing irrigation.   The crop of wheat stood about 4 inches high ; the field was divided by bourrelets into planes the same as was the case on the French arable lands.   It took 19 minutes for the water to run the length of the plane, which was 45 metres (49 yards) and 5 metres (about $16\frac{1}{2}$ feet) broad, that is 19 minutes to irrigate 2·45 ares or under similar circumstances the field was irrigated at the rate of 14 hours per hectare (5 hours per acre) ; but very often when one plane is half irrigated and there is plenty of water the second is started, and so on, reducing the time by nearly one half. This irrigation was superintended by a couple of boys, the elder of whom was 14 years of age ; both were brothers of the proprietor and received no other help. They made use of a special irrigating spade with a short curved handle, giving it the appearance of a large but short-handled hoe.

As at St. Remy few or no vannes were to be seen in the secondary distributors, the water being arrested and diverted on to the field by means of a board or large stone which blocks its way, any little crevisses being plugged with grass, earth, &c. On a neighboring field which had been irrigated the morning previous to our visit the water had penetrated to a depth of 8 to 9 inches. If there is plenty of water more than one plane is irrigated at once, even 3 to 4 at a time. It being winter water was in less general demand and the peasants were less careful in its application and wasted much.

On some of the fields here the bourrelets were occasionally branched more or less as appears in Fig. VII. These branches are made in order to counteract any unevenness in the slope of the field which often causes the water to run close alongside the bourrelet instead of spreading uniformly over the plane.

Some planes here were 120 metres (nearly 6 chains) broad and 150 metres (7½ chains) long.

On a plane 4½ metres (14½ feet) broad and 97 metres (4 chains) long, the depth of water covering the soil as it flowed along, varied from 1 to 3 centimetres. The irrigator told us that with planes like this he could do 4½ hectares (11 acres) in a day of 24 hours. At another place we were told it takes an hour to irrigate ½ hectare (1⅓ acres) equal to 32 acres in 24 hours, which statement judging by the answers to our frequent enquiries elsewhere appears very much exaggerated. Even the former big result can only be obtained when several planes are irrigated at one and the same time.

The velocity of the water in the distributors was very variable; thus a length of 15 metres (57 inches) required 30, 32 and 35 seconds, and 100 metres (328 feet) required 3 minutes 20 seconds for the water to run from end to end. In a distributor 45 centimetres broad and 21 to 32·5 centimetres deep the water required 22 seconds to flow a length of 13 metres, 40 seconds for a length of 19 metres, and 45 seconds for a length of 23 metres. In the Canal itself the water flowed at the rate of 3 minutes for 80 metres (262 feet). In a conduit with a mean breadth of 85 centimetres, and the water 17 centimetres deep, the water flowed a length 23 metres in 33 seconds, equal to a delivery of about 101 Litres (22¼ gallons or 3·53 cubic feet) per second. In another conduit with 80 centimetres

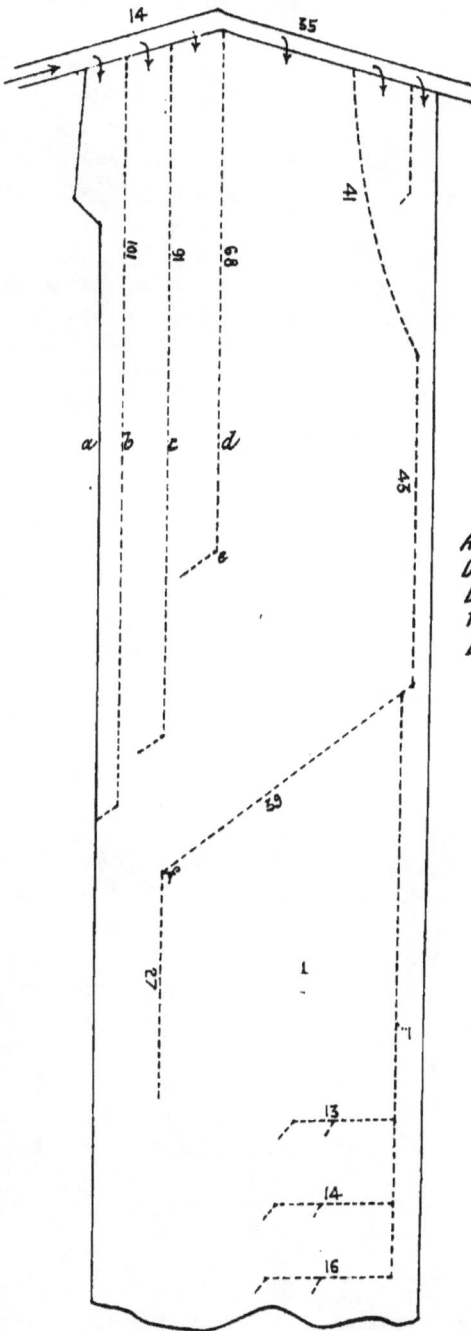

Plate V

Fig 8

A Wheat field near Granada
Direction of water denoted by arrows
Dotted lines denote bourrelets
The figures indicate lengths in metres
Distances between bourrelets
     a & b is 3 metres
     b . c : 5 "
     c . d : 6 "
     e . f . 19 "

mean breadth, a depth of water varying from 22·5, 26, 28, 23, to 25 centimetres (mean 25) the water took 43 and 45 seconds to flow a distance of 20 metres, equal to a delivery of 93 and 89 Litres (20 to 19 gallons) per second respectively. We found variations like the above in velocity and delivery wherever we took measurements.

From the President of the Board of Agriculture at Valentia we learnt that the peasants try to reduce the slope of the field as much as possible. If the water flows over the field too fast and washes the soil they level more; but the peasants, without being able to explain why, know at once by sight when anything is going wrong. The length of the plane depends on the slope. The more level the plane the greater the control of the irrigator over the water. He says he believes many arable fields are perfectly level, the water being forced along by the fall from the conduit into the distributor. Most of the conduits and distributors were laid out years ago, when the main canal was built; he does not know that their relative distances are in any way limited, it all depends on the conformation of the ground. The peasants appear to have laid out the secondary and lesser distributors in the same way as already described of their French brethren.

To cereal crops water is given; 1st time : either before or after sowing, in December or January. 2nd time : in February. 3rd time : in April, between the formation of the ears and the time of

flowering. 4th time: in May, after flowering.
The latter is considered an important time to irrigate, as upon this irrigation depends the proper
maturing of the grain. At this time it is sometimes
difficult to tell whether the ground possesses sufficient
moisture to ripen the grain without creating a
shrivelled produce, and where hot winds prevail
only long experience can determine this.

In winter one can irrigate at all hours, just when
it is most convenient; but from May onwards, until
the sun's heat again decreases, any irrigation for
cereals must be carried on in the night-time.

When water runs short the canal guardians give
to those fields which most require it.

Some planes which we measured on land under
wheat, near the village of Las Casas de Barcenas,
gave the following results: one plane, 173 metres
(8½ chains) long, had a slope of 0·0037; another
plane, 140 metres (about 7 chains) long, had a slope
of 0·0048; and a third, 30 metres (1½ chains) long,
had a slope of 0·0034.

At Grenada we found by measurement that the
distances between conduits and between secondary
and lesser distributors were not fixed by any rule,
but were placed where required in accordance with
the natural matured slope of the land, and little
affected by the size of the holdings; the distances
apart of these distributors varied from a few metres
with the smallest to 140, 182, and on one occasion
290 metres with secondary distributors and conduits.

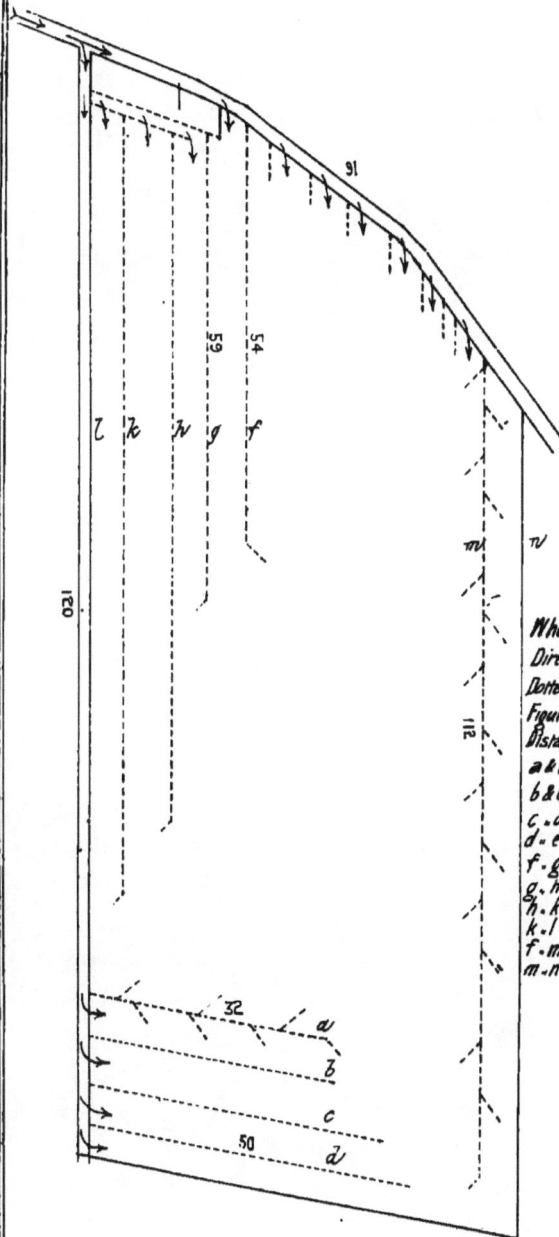

Plate VI

Fig 9

Wheat field near Granada
Direction of water denoted by arrows
Dotted lines denote the bourrelets
Figures indicate lengths in metres
Distances between bourrelets

a & b	is 5	metres
b & c	" 6	"
c - d	" 5	"
d - e	" 4	
f - g	" 6	"
g - h	" 4	"
h - k	" 8	"
k - l	" 5	"
f - m	" 38	"
m - n	" 5	"

Figs. VIII. and IX. describe the irrigation of arable land, as practised here, and need no further comment.

In a canal here the water flowed very sluggishly, having a velocity a little above one foot and a half per second (it flowed 73 metres in two minutes thirty seconds), and it should vary from two to three and more feet per second; in a secondary distributor (from this canal below a considerable fall) 40 centimetres mean breadth, and 53 centimetres depth of water, the velocity was 31 metres in 29 seconds ($3\frac{1}{2}$ feet per second), equal to 226 litres (8 cubic feet) per second.

The works were not in the same good order as at Valentia and in France; the sandy nature of the soil may have something to do this. While in other parts of Spain, as we were informed, the peasants were improving, and took good care of their irrigation works; the Andalusians make little or no progress.

They irrigate here at the commencement of November, in April before the flowering of the cereals, and about May after flowering; therefore the principle is the same as at Valentia and in the south of France.

At Grenada, above the Alhambra, we visited a small conduit cut partly in the rock and partly built of stone. Its course was along the mountain side, and it brought water a distance of three to four miles from a small mountain torrent to irrigate the

summer gardens of a Spanish Grandee. This was
the only acqueduct on a mountain side we saw, but
we have reason to believe from what the peasants
told us that they are tolerably common. See Fig. X.

They make use of a peculiar instrument here for
working the soil; it has the body of a primitive
plough, but instead of share and mould-board it is
fitted with a pointed iron head about 12 inches long.
It works to a depth of 4 to 6 inches. In one place
where the peasants were levelling a small strip of
land, shovels were used, and the earth removed in
orange baskets by hand. We saw soil being removed
this way several times in Provence; it was chiefly done
to make bare rocks, and elevated situations generally,
fertile. Nearly all transportation of agricultural
produce is carried on by means of pack-mules and
horses, although the country roads are very fair.
(At Gerona large forks were used for tilling the
soil. Much guano appeared to be used, but the
wheat-crops do not seem to exceed $3\frac{1}{2}$ quarters
per acre).

Near Seville, notwithstanding the terrible heat
and want of rain there are no irrigation works; but
watering is carried on to a certain extent on a small
scale by means of " norias," *i.e.* wells fitted with a
very primitive chain pump and gear worked by cattle.

Near Saragossa there is a good deal of " terrace
irrigation." In this system the water having irrigated one terrace, collects and irrigates a second one
lying immediately below it and so on, as exemplified

Plate VII

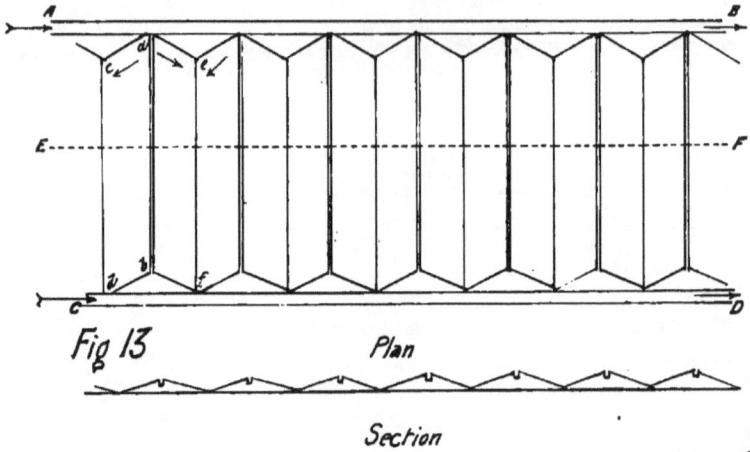

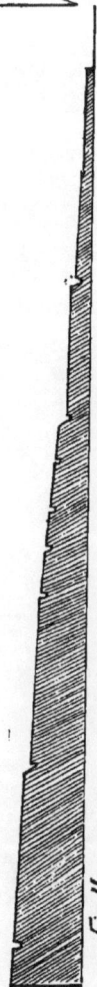

Fig 13   Plan

Section

Fig 10

Small conduit, on the mountain side, above
Granada, partly cut in the rock, & partly
built of stone

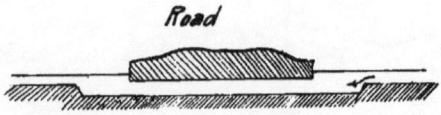

Road

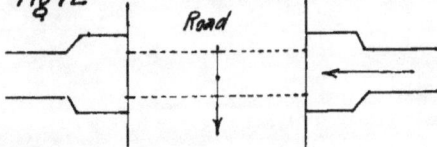

Fig 12   Road

Plan & section of a canal crossed by a road, on the same, or
nearly the same level.

Fig 11 Section of Terrace Irrigation (various Sub-systems combined.

in Fig. XI. In some parts of Germany this system is, I believe, very wide spread. It is called there "Hang-Bau." Sometimes the terraces are perfectly level, at others they have a considerable slope.

Occasionally canals and distributors have to cross the roads; so as not to interfere with the traffic, the canals are built under the road, as sketched in Fig. XII. At Saragossa such a canal crosses under the railway lines.

The country is extremely subdivided, the separate holdings are consequently very small, and one frequently hears that a single olive-tree is owned by two or more proprietors. It is extremely rare to see a hedge or fence, but in Andalusia aloes and cacti are grown as hedges, and make very formidable fences.

III. MILAN AND LUCCA (ITALY) AND DOUAI (FRANCE).

In Italy we visited only a few irrigated fields. Near Milan some meadows are irrigated on the bed-work system. The land is carefully surveyed and laid out by experts, as exemplified in Fig. XIII. The water is diverted from the conduit A B into the distributor a b from which it overflows into the drains c d and e f respectively, and thence into the main drain C D, which as often as not serves as a conduit for a meadow situated at a lower level. It is costly in the first out-lay, but in the long run highly remunerative. The slope of the sides of these beds is about 0·5. In one locality here the water, obtained from a spring, is in winter several degrees warmer than the atmosphere, hence so long as it flows over the meadow only those blades of grass which stretch out above the water get frost-bitten, the rest of the grass thriving amazingly.

When there is no more danger of frost the water is diverted into another channel. This system is by no means merely intended to keep off the frost, being in very common use for irrigating land laid down to grass.

One small meadow we visited was watered by

being totally submerged, the water was about four inches deep on the surface, it flowed in at one corner and found an outlet at the opposite end. In some cases this system is resorted to in order to fertilise arable land, especially when the irrigating waters are turbid; the water is then retained on the field until the greater part of the ooze has been deposited, after which it is run off; the grass on these inundated meadows, however, is said to be inferior to that irrigated in the ordinary way.

At Lucca a common practice with arable land is to plough deep furrows between the rows, or every second and third row of the crop, drawing the furrows very straight; then when irrigation becomes necessary the water is allowed to enter these furrows from the distributor at the top of the field, and flows along them until it reaches the bottom of the field as is shown in Fig. XIV. This field was about 10 chains long; the furrows about 18 inches apart. The straight lines indicate the furrows, the letters A B the conduit, and C D the distributor; the water flowed from the distributor down two to four furrows at one and the same time.

There was no appreciable slope on most of the fields irrigated on this system.

On the Arno, at Florence, there is to be seen a weir thrown across the bed of the river to raise water required for factory purposes. Fig. XV. gives a tolerable idea as to how it is managed. In case of excess of water the sluices at C can be opened, and inundation due to retention of the excess of water by the weir prevented.

On visiting the farm attached to a large beet sugar-mill at Masny, near Douai, in the north of France, we were shown a field which had been laid out for irrigation as described by Fig. XVI. The field was almost square, and contained 23 hectares (57 acres). A permanent brick canal, 9 inches wide, ran along one side and along half of the adjoining side. When the field was to be irrigated a deep furrow was temporarily cut along the back, as it were, of the field, indicated by the dotted lines B C, and from this deep furrow the water flowed down the furrows, generally several at a time, between the rows of beet. The superfluous water flowed out at D through the ditch. Several other beet fields were irrigated, but all the distributors were of temporary construction, very much on the system explained by Fig. XIV. at Lucca.

Plate VIII

A   B
C   D

Plan

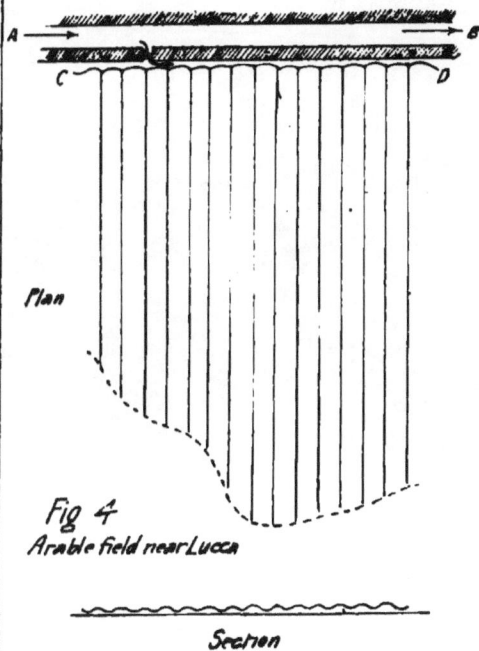

Fig 4
Arable field near Lucca

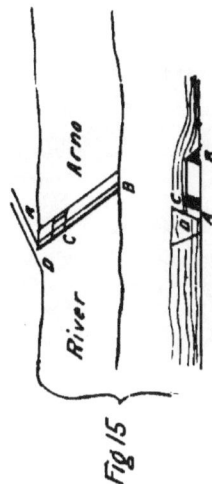

Section

Arno

River

Fig 15

A B The weir. C The Sluice D The Conduit

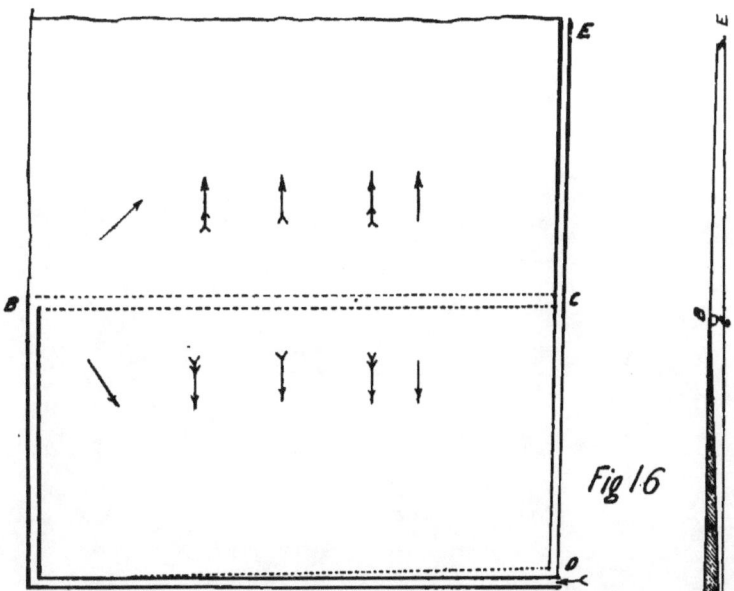

Fig 16

A B. Permanent brick canal, 9 in. wide carrying water from a well, to BC the temporary distributor
The arrow → indicates the slope of the land ⇢ flow of water ⇢ direction of the beet rows, in
the furrows of which the water flows D E is the open drain, with outflow at E

Plate VII.

THE MAIN CONCLUSIONS ARRIVED
AT ARE:

1. That all land is fitted for irrigation if attainable by irrigation works—the position of the land itself being a secondary consideration. Mountain slopes, hilly spurs, valleys are all alike brought into cultivation as soon as water is obtainable to irrigate them; at times, as we have shown, soil is even transported to elevated and barren localities where water, having been brought by canals, has become available. Hence irrigation extends the area under cultivation and increases the produce of the country.

2. That irrigation is more suited to small or peasant proprietorship than to large holdings, even than to holdings of 100 acres. The reason for this is that the irrigation of land requires a care, attention, and scrupulous cleanliness, which, while they repay manifold, are nevertheless of too minute and important a character to be left in the hands of a paid labourer. It requires the immediate supervision and direct work of the proprietor himself, who alone is sufficiently stimulated by the return he receives. It is all very well to say that it must be a despicable man who working by the side of his master allows that master to work harder than he himself does, yet it is expecting too much from any ordinary man

that he should care, seeing he is not impelled by the same hope of proportionate reward.

3. That any man who understands anything of agriculture even in its rudest form is capable of laying out the lower systems (Figs. I. and III.) of meadow irrigation. For arable-land irrigation more knowledge and experience are required.

4. That the size of the fields to be irrigated depends on the configuration of the land, even (though to a far lesser extent) with the higher system of irrigation (Fig. XIII). Even on a plain the natural unevenness of the surface is general, hence the size of the fields is in so far limited. Perhaps, however, were the proprietorship of the land not so excessively subdivided the fields would in many cases be very much greater; notwithstanding this the greater part of the fields are large enough to admit of bullock labour. As far as we were able to judge it appeared as though fully one-third of the cultivation of the soil was carried on by manual labour only.

5. That irrigation forms a payable investment, both to the capitalist who invests his money in the irrigation works, and to the landed proprietor who avails himself of the water, and who would otherwise have no crops or none at least to speak of.

6. That the produce per acre in those districts is not so great as it might be, if a knowledge of the use of artificial manures were more wide-spread amongst the cultivators of the land.

FINIS.

TABLE OF MEASURES QUOTED.

——————◆——————

(Extracted from Molesworth's Pocket Book).

1 Metre = 3·28 feet; 1 square metre = 1·196 sq. yards.
1 Are = 100 sq. metres = 1076·4 sq. feet.
1 Hectare = 10,000 sq. metres = 11,960 sq. yds. = 2·4711 acres.
1 Litre = 0·22 gallons; 1000 Litres = 35·317 cubic feet.
1 Hectolitre = 2·751 bushels = 22 galls. = 3 53 c. ft. = 100 Litres.
1 Chain (or 22 yards) = 20·116 metres.

EXTRACT FROM " CYPRUS, AS I SAW IT IN 1879."
By Sir S. Baker.

Page 34—" We now arrived at the spot where the water is led by a
subterranean aqueduct to Larnaca. . . .

" In a search for water springs the Cypriste is most intelligent, and the
talent appears to be hereditary. If a well is sucessful at an elevation
that will enable the water to command lower levels at a distance, it may
be easily understood that the supply of one well representing a unit must
be limited. The Cypriste well-sinker works upon a principle of simple
multiplication. If one well produces a certain flow, ten wells will multiply
the volume, if connected by a subterranean tunnel, and provided the
supply of water in the spring is unlimited.

" It appears that Cyprus exhibits an anomaly in the peculiarity of a
small rainfall, but great subterranean water-power ; some stratum that
is impervious retains the water at depths varying according to local conditions. The well-sinker commences by boring, or rather digging, a cir-
cular hole two feet six inches in diameter. . . ."

Page 35—" The rapidity of the well-sinking naturally depends upon
the quality of the soil ; if rock is to be cut through it is worked with a
mason's axe and a cold chisel. Fortunately the geological formation is
principally sedimentary limestone, which offers no great resistance. At
length the water is reached. The well is now left open for a few days
that an opinion may be formed of the power ; if favorable another precisely similar well is sunk at a distance of fifteen or sixteen yards in the
direction towards the point required by the future aqueduct. The
spring being satisfactory the work proceeds with vigor. We will accept
the first well as forty feet in depth ; if the surface of the earth were
an exact level, the next well would be an equal depth ; but as the water
retains its natural level, the vertical measurement of each shaft will
depend on the formation of the upper ground. The object of the well-
sinker is to create a chain of wells united by a subterranean tunnel, in
order to multiply the power of a unit and to obtain the entire supply of
water ; he therefore sinks perhaps ten or twenty wells to the same level,
and he cuts a narrow tunnel from one to the other, thus connecting his
shafts at the water-line, so as to form a canal or aqueduct. Precisely as
the mole upheaves at certain intervals the earth it has scraped from its
gallery, the well-sinker cleans his tunnel by sending up the contents
through the vertical shafts fifteen yards apart around the mouth of which
a funnel-shaped mound is formed by the debris.

" These preliminary walls being completed and the water-volume tested, the neighborhood is examined with the hope of discovering other springs that may upon the same principle be conducted towards the main line of the proposed aqueduct. It is not uncommon to find several chains of wells converging from different localities to the desired water-head, and as these are at higher levels a considerable hydraulic power is obtained, sufficient in many instances not only to fill the tunnels but to force the water to a greater elevation if required.

" The water-head being thoroughly established, the sinking of a chain of wells proceeds, and the tunnels are arranged at a given inclination to conduct the water to the destined spot. This may be many miles distant, necessitating many hundred wells, which may comprise great superficial changes; hills that are bored through necessitate deep shafts, and valleys must be spanned by aqueducts of masonry. . . ."

Page 92.—Here the author describes the carelessness of the irrigators on Mr. Mattei's farm, Mr. Mattei being one of the largest landed proprietors in Cyprus.—Nowhere else does Sir Samuel notice waste of water in irrigation, although he frequently throughout his book refers to irrigation as carried out by the peasants on their own lands. He consequently would support my conclusion No. 2.

[I have inserted the above as it contains information not to be found in any handbook.—H. L. R.]

REVIEWS OF REPORT ON THE SUGAR INDUSTRY IN QUEENSLAND.

"The continued progress of the Sugar Industry in Queensland prompted some capitalists to commission Mr. Roth to visit the country for the purpose of investigating and reporting on this one, of its important sources of wealth; and the report, which is now published in a handy book form, shows how eminently judicious was the selection of this gentlemen for such a purpose."—*Brisbane Courier*, June 25, 1880.

"The whole book is a partly mournful and partly ludicrous testimony to the fact that Mr. Roth knows as much of the subject he is writing on as he knows of the origin of the creation of man. It is a remarkable specimen of that false knowledge which is far more dangerous and mischievous than the densest Boeotian ignorance."—*A Queensland Paper.*

"The subject appears to have been very ably handled by Mr. Roth." —*St. George Standard*, 3rd July, 1880.

"It promises a dood deal, and really gives very little that is trustworthy; because, on close examination, its tone and tenor destroy confidence in the ability or inclination of the author to give reliable information on the subject on which he treats."—*Brisbane Telegraph*, June 28, 1880. (*Second Review*).

"It is a book of 109 pages, besides maps, appendices, and index—all thoroughly done, and perhaps the most reliable, as it certainly is the most compact document yet given on the question."—*Sydney Morning Herald*, July, 1880.

"The report contains much valuable information on the Sugar-growing districts in Queensland. It is furnished with tables of rainfall at some of the chief Sugar plantations in the colony, as well as with useful information as to the peculiar difficulties of Sugar-growing in Queensland. There are some interesting remarks on the history of the rise of the Queensland Sugar trade."—*The Statist*, September 11, 1880.

" . . . By pointing out the advantages and disadvantages which the present growers and manufacturers possess and have to contend against, Mr. Roth has laid the question fairly and clearly out in his report."—*The European Mail*, September 24, 1880.

"In the compass of about 120 pages we have here a model volume containing a vast amount of valuable information on the production of Sugar in Australia. . . . . . . After describing successively the

Sugar-growing districts, their geology and soils, and the rainfall, the origin and progress of the Sugar industry is detailed, the question of foreign labor discussed, central factories and small mills described, average yield per acre and cost of production given, the distillation of rum, and the colonial tariffs on Sugar noticed. Lastly, the future yields and prospects of culture are gone into and the local consumption of Sugar given. . . . . . . We heartily commend this little volume to all interested in Sugar cultivation, for its facts and figures may be studied with advantage."—*Journal of Applied Science,* October 1, 1880.

AGRICULTURE AND PEASANTRY OF EASTERN RUSSIA, BY H. L. ROTH.

From the *Daily News,* September 14th, 1878.

A SKETCH OF THE AGRICULTURE AND PEASANTRY OF EASTERN RUSSIA, BY H. L. ROTH (*Ballière, Tyndal & Co.*), comprises in comparatively small compass a great deal of information regarding the agricultural systems, the rural economy, and social institutions of the portion of European Russia lying between the famous commercial city of Nishni-Novgorod and the northern head of the Caspian Sea. In this district, or, to speak more exactly, in the province of Samara, the author very recently spent two years, devoting that time to practical farming in the neighborhood of Timashevo, a village situated about 60 miles east of Samara town, save such of it as was spent in travelling through the neighbouring country for a few months, visiting what districts appeared to be interesting from the agriculturist's point of view. The author's method is somewhat like that of Arthur Young's—that is to say, he mingles his own personal observations with statistical and other particulars obtained on the spot. The details, however, regarding soil, crops, course of husbandry, stock, agricultural machinery, and implements in use, and the communial government and distribution of lands, bear the impress of actual experience. Mr. Roth's volume contains some curious particulars regarding the cattle plague in the steppes, popularly reported to be its origin and home. His book is strictly practical, and is studiously plain and straight forward in its mode of conveying information. The author is, however far from being deficient in literary power, as will be seen in his introductory chapter on the general features of the country, and above all in his vigorous and picturesque sketch of travelling in the roadless and almost trackless plains which are characteristic of this portion of the Russian empire,

www.ingramcontent.com/pod-product-compliance
Lightning Source LLC
Chambersburg PA
CBHW022017190326
41519CB00010B/1551